1 MONTH OF
FREE
READING

at
www.ForgottenBooks.com

By purchasing this book you are eligible for one month membership to ForgottenBooks.com, giving you unlimited access to our entire collection of over 1,000,000 titles via our web site and mobile apps.

To claim your free month visit:

www.forgottenbooks.com/free912491

ISBN 978-0-266-93703-6
PIBN 10912491

This book is a reproduction of an important historical work. Forgotten Books uses
state-of-the-art technology to digitally reconstruct the work, preserving the original format
whilst repairing imperfections present in the aged copy. In rare cases, an imperfection in
the original, such as a blemish or missing page, may be replicated in our edition. We do,
however, repair the vast majority of imperfections successfully; any imperfections that
remain are intentionally left to preserve the state of such historical works.

Control of Decay

in Bolts and Logs of Northern Hardwoods

during Storage

Theodore C. Scheffer
J. W. Jones

Station Paper No. 63

Northeastern Forest Experiment Station

Upper Darby, Pennsylvania
Ralph W. Marquis, Director

1953

United States Department of Agriculture • Forest Service

FOREWORD

Many wood-using plants in the Northeast store large quantities of hardwood logs for rather long periods. Sometimes a large volume of the wood is spoiled by decay during the storage period. A number of people have asked: "How can we prevent this loss?"

To answer this question, the Northeastern Forest Experiment Station stimulated a cooperative study to gather the best available information on control of decay during storage, and to make it available to wood-using plants. This report is the result of the study.

As the authors point out, further testing is needed for some of the treatments discussed. This report is certainly not the final word on the subject; but it may be useful as a guide until the results of further study become available.

Ralph W. Marquis
Director

Control of Decay

n Bolts and Logs of Northern Hardwoods

during Storage

by

Theodore C. Scheffer *and* T. W. Jones[1]
Pathologists

Division of Forest Pathology
Bureau of Plant Industry, Soils, and Agricultural Engineering
United States Department of Agriculture

ONE OF THE PROBLEMS of many wood-products manufacturers who use northern hardwoods is the prevention of serious decay in their stored bolts or logs during warm weather. The difficulty arises because a substantial portion of the wood milled in the summer and early fall is cut several months ahead of time and consequently is stored for long periods during which temperatures are favorable for decay fungi.

At present, means of adequately reducing this lag between cutting and use seem to be but partly feasible. With some products, it is possible to convert the logs promptly into boards or blanks that can be air-dried comparatively fast and then stored under cover. However, this procedure

[1]Mr. Scheffer is stationed at Madison, Wis., in cooperation with the Forest Products Laboratory, which the U.S. Forest Service maintains in cooperation with the University of Wisconsin. Mr. Jones is stationed at New Haven, Conn., in cooperation with Yale University.

seems to have limited application—largely for economic reasons.

Therefore attention has centered on the possibility of protecting bolts and logs by storage methods or by surface treating them with chemicals designed to keep out decay-producing fungi. Results of limited work on the problem indicate that much may be accomplished in these ways. The following information contains some of the more pertinent evidence on these points and such practical guidance as appears warranted from the information at hand.

Suggested protective measures are considered to be applicable in the United States throughout the range of northern hardwoods from Maine westward to Minnesota and extending southward to about the 44th parallel.

WATER STORAGE

Northern hardwood logs can be kept practically free of decay by storing them in fresh water shortly after cutting. This is known from general experience with water storage, and it has been specifically demonstrated in a test made on yellow birch.[2] Yellow birch veneer bolts placed in a pond incurred no decay during a storage period of 9 months, from February through October; but comparable bolts stored in ricks were decayed an average distance of 8 inches inward from each end.

The marked protection given by water storage results from the oxygen-limiting effect of the water. This is sufficient to inactivate most fungi that come in contact with the wood. It follows that complete submersion is needed for maximum protection; however, a log that floats comparatively low will be almost as well protected. It has been estimated[3] that about 40 percent of woods-run northern hardwood logs will sink promptly and the remainder will either sink after a time or else float low enough to virtually preclude decay.

Of course the maximum benefit from water storage will be had if the logs are put in water before significant damage by decay can occur. Wood cut in the fall and winter

[2]The test was made at Laconia, N. H., in cooperation with Tekwood, Inc., and the Northeastern Forest Experiment Station.

[3]U. S. Forest Products Laboratory. Pertinent facts on salvage of New England timber. Forest Prod. Lab. release R1183 (mimeo.). 1938.

should be put in the water by late April or early May.   An acre of water about 4 feet deep reportedly may accomodate about 500,000 board feet of logs.   In some situations it may be advantageous to do as much hauling as possible while ice is thick enough to permit unloading directly on the storage pond.

Water storage presumably would be most feasible with logs, although in certain situations it probably could also be used with smaller material such as turning bolts. In any case, it is limited to locations where there is ample fresh water. Brackish water is unsafe because of the accompanying danger of marine borers.

A natural body of water is not necessarily a require-ment;   shallow ponds can be constructed where the upper soil is loose enough to be moved and the subsoil is impervious enough to prevent rapid seepage.   Recovery of "sinkers" is an item of cost that can be held to a reasonable level by using comparatively shallow water.   ·

A large hardwood veneer mill in New England has con-structed long, narrow, clay-lined pits with a capacity of 2 million board feet of logs.   The pits are about 20 feet wide, and logs are stacked in them crossways and covered with water pumped from an adjacent. river.   Logs may be re-moved by grappling or by lowering the water·level. Costs of such storage are said not to exceed those of dry decking, and the original installation is comparatively inexpensive. Details will be supplied upon request.

Except over very long periods, storage in water will not cause the wood to become objectionably discolored. Water marking usually will occur only near the log surfaces.

S T O R A G E   U N D E R   W A T E R   S P R A Y

The question is sometimes asked whether decked logs can be similarly protected by water applied as a continuous spray.   Attempts along this line have been made in connec-tion with large conical piles of pulpwood. Results were not sufficiently good, however, and the method apparently is little used at the present time.

It is probable that the ineffectiveness of spraying can be attributed to failure to get all portions of the storage pile sufficiently wet.   Excellent results recently were obtained by a New Hampshire concern in commercial trials on yellow birch logs, utilizing rotary sprays of the pulsation-actuated type. If the economics of water-spraying are attractive to a manufacturer, the method has sufficient promise to warrant additional trials.

STORAGE UNDER WET HAY

Decay has been materially reduced in turning bolts of hard and soft maples, white and yellow birch, and beech, by storing them on end during the winter under a covering of hay that is wet down from time to time during subsequent warm, dry weather (fig. 1).

The hay, acting as a heat insulator and sunshade, prolongs the period during which the wood remains unfavorably cold for decay. Wetting of the hay helps keep the bolts cool. Excess water from the wetting, supplemented by rain water, keeps the ground quite wet, and this situation, coupled with cool air, apparently can give the lower ends of the bolts a high degree of protection. In a large number of bolts observed after storage under hay, the decay was confined almost entirely to the upper ends.

From limited production records and personal observation, it is estimated that decay losses in northern hardwood turning bolts stored through an entire warm season under wet hay commonly amount to about 20 percent and sometimes may approach 30 percent. Although such losses are large, they nevertheless are only about half of what often is incurred in the same period with bolts stored in conventional ricks.

If the upper ends of the bolts stored under hay were also sprayed with a good fungicidal chemical, it is possible that the decay could be reduced considerably more. Such additional protection would be fairly inexpensive and seems to offer considerable promise because seasoning checks in the upper ends of the bolts under wetted hay are not prevalent. The significance of seasoning checks in connection with chemical control of decay is discussed in the following section.

CHEMICAL TREATMENT

Logs of hard and soft maple and of white and yellow birch can be protected for moderate to comparatively long periods by treating the ends properly with chemicals that prevent entrance of decay fungi.

Treatment of the bark is unnecessary because of its natural fungus resistance. Likewise, it appears from limit-

4

Figure 1.--Turning bolts of northern hardwoods stored under intermittently wetted hay. This treatment reduces decay materially.

Table 1.—Solutions used for chemical treatment of logs

Chemical	Suggested amount per 100 gallons of water
	Pounds
Dowicide G (sodium pentachlorophenate)	21
Melsan (ethylmercuric phosphate + sodium pentachlorophenate)	12
Noxtane (sodium pentachlorophenate + borax and soda)	30
Permatox 10-S (sodium pentachlorophenate + borax	30
Santobrite (sodium pentachlorophenate)	21

Table 2.—Decay in northern hardwood turning bolts after prompt spray treatment of the ends with a 5-percent solution of sodium pentachlorophenate

Species	Month cut and treated	Storage period, ending in October	Treated bolts examined	Average depth of decay from end surfaces	
				Treated bolts	Untreated bolts
		Months	Number	Inches	Inches
White birch	November	11	40	12	13
	February	8	40	11	[1](14)
	April	6	37	13	15
	June	4	37	13	17
Yellow birch, hard maple, and soft maple	February	8	6	15	[1](14)
	April	6	10	13	[1](14)
	June	4	17	3	[1](14)

[1]Decay penetration was not actually measured, but was estimated from the general condition of representative bolts to be near the maximum average found in the treated bolts. Penetration measurements were not made in all of the untreated control groups because decay obviously was so extensive that the cost of sawing out boards for the purpose did not appear warranted.

ed evidence that exposed knots and ordinary barked patches along the sides may be left untreated without serious disadvantage. By not treating the sides, it is possible to treat after the logs have been put in the storage pile. However, it is important not to delay treating too long.

To be effective, treating obviously must be done before fungus infection occurs. In warm weather, this may require treatment within 48 hours after cutting. In cold weather the interval between cutting and treating may safely be extended to a much longer time. Specific suggestions in this respect are given later.

Fungicide Spray

The simplest form of chemical treatment consists of spraying with an appropriate fungicide solution. All that is necessary is that the log or bolt ends be thoroughly wetted. A garden-type sprayer with a capacity of 3 to 4 gallons is suitable for this.

Chemicals and concentrations considered adequate for general use on logs are listed in alphabetical order in table 1. When concentrated, these chemicals may act as a skin irritant, particularly to persons allergic to them. So the spraying should be done with care to avoid breathing droplets of solution and to avoid accumulations of the solution on exposed parts of the body.

In northern hardwood areas, spray treatment by itself is suitable for use only during the warmer time of the year, beginning in April, and for storage periods that are comparatively short, probably no longer than about 2 to 3 months. For longer storage periods the fungicidal proteotion needs to be supplemented by some measure to retard the development of seasoning checks in the exposed ends. These seasoning checks create numerous breaks in the barrier of treated wood, through which infection can occur (fig. 2).

Evidence of what may be accomplished by spray treatment shortly after cutting is presented in table 2. This gives results of a test[4] made in Maine on turning bolts of mixed hardwoods stored in conventional ricks. Comparable results would be expected in material of log size. Extent of decay was observed on slats sawed out of the bolts. Decay during 4 months of summer storage in bolts of yellow birch and maple, (but not of white birch) was markedly reduced by treating. The poor protection of the white birch is ascribed to earlier end-checking in this species than in the others. None of the bolts cut prior to June, and stored for a minimum of 6 months, were benefited by the treating.

---

[4]The test was made with the cooperation of Penley Brothers, West Paris, Maine; the U. S. Forest Service; and the University of Maine.

Figure 2.--Typical severe end-checking in white
birch bolts. Seasoning checks permitted decay fungi to enter the bolts despite the
preservative solution applied to the ends by
spraying.

Logs and bolts can be protected for much longer periods if the fungicidal spraying is followed by a suitable coating to retard end-checking. Good control of decay was obtained with such a combination in white birch bolts for as long as a year in a test made in northern Wisconsin.

Nearly complete control was obtained with certain of the coatings applied in double thickness; consequently, it is suggested that a double coating be considered if maximum protection is wanted and the storage is to be in open sunlight for a period in excess of 6 months. A single coating appears sufficient in partial shade. With the best of treating, however, it is unlikely that logs can be successfully stored through more than one warm season (i.e., spring through fall) because of infection ultimately introduced through the sides by bark- and wood-boring beetles.

Asphaltic preparations have been the most effective of the coating materials tested in connection with long storage periods. These are of two types, emulsions and cutbacks. The emulsions are somewhat easier to apply, but they are suitable only when temperature of the wood is above 40 degrees; in single application, they also seem to be slightly less effective than the cutbacks over a long period. Although the cutbacks can be applied during the winter, application is done most effectively by warming the material or by using it on days that are warm enough to keep the material reasonably spreadable.

In applying an end-coating, it is essential that sufficient material be deposited to provide, when dry, an unbroken cover. This can be accomplished readily if the coating material has the consistency of very thick paint or of light grease. If two coats are applied, the consistency may be somewhat thinner than if one coat is to be depended on. Twenty-four hours seems to be a long enough interval between coats. The log ends should be covered to the outer edge of the bark to retard moisture loss from the inner edge of the bark as well as from the exposed wood. The spreading can be done fairly rapidly with a paddle or stiff brush. Spray application with high-pressure equipment also may be feasible.

Results of the aforementioned test in which both a fungicidal spray and end-coating were used are given in table 3. The depths of decay penetration in the Wisconsin test were ascertained by splitting the bolts. Since the decay data pertain to individual bolt ends, penetrations

Table 3.--Decay in 7-inch (mean diameter) white birch bolts after prompt spray treatment of the ends with a 5-percent solution of sodium pentachlorophenate followed by a coating to retard checking

Place and storage period	Bolts treated	End-coating[1] used	Depth of decay from end surfaces at end of test	
			Mean maximum	Mean average
	Number		Inches	Inches
West Paris, Me. (June to October 4 months)	40	Asphalt emulsion (C-13-E)	7.0	2.0
	40	None; untreated controls	23.0	17.0
Oneida Co., Wis. (June to June 12 months)	10	Asphalt emulsion (C-13-E)	6.0	3.2
	10	Same, 2 coats	1.5	.5
	10	Asphalt emulsion (C-13-HPC)	7.0	3.4
	10	Same, 2 coats	.3	.1
	10	Asphalt cutback (Semi-Mastic 214)	5.5	1.7
	10	Same, 2 coats	.3	.1
	10	Asphalt cutback + clay filler	3.9	1.5
	10	Same, 2 coats	4.4	.6
	20	None; untreated controls	[2]12.0	10.0

[1]Coatings from Flintkote Company.

[2]Bolts for this test were only 2 feet long; so decay 12 inches deep denotes penetration to the center of the bolt.

Table 4.--Decay in 15½-inch (mean diameter) yellow birch veneer bolts during 8 months of water storage or of dry storage with protection by spraying the ends with 5-percent sodium pentachlorophenate and following with a coating of asphalt cutback[1]

Treatment	Bolts treated	Depth of decay from end surfaces at end of test	
		Mean maximum	Mean average
	Number	Inches	Inches
Stored in water	10	0	0
End-treated promptly after cutting	10	7.5	0.6
Same, except patches of bark were first removed to simulate effect of rough handling; barked areas sprayed but not coated	10	8.6	0.4
End-treated 2 months after cutting	10	9.6	2.2
None (controls)	10	16.2	8.0

[1]All bolts were cut in early February and the decay observations were made in early October.

listed in table 3 are just one-half the average amounts present in an entire bolt.   Consequently, to eliminate the major portion of decay (indicated by the mean average depth) in the untreated bolts, it would be necessary to trim the bolts by 20 to 34 inches.   In contrast, the most effectively treated bolts would be shortened by less than $\frac{1}{2}$ inch, an amount that normally would be trimmed off even if there were no decay.   The difference in required amounts of trimming would be still larger if manufacturing requirements called for elimination of decay streaks deeper than the main zone of infection (designated mean maximum depth in table 3).

Results of a test[5] of the combination of spray treatment and end-coating applied in cold weather and on large yellow birch veneer bolts are presented in table 4.   Results obtained in the same test with water storage, already referred to, are included.

The chemically treated bolts were stored in ordinary manner on the mill yard.   The major observations of decay were made on sheets of veneer cut from the bolts.   Additional variables considered in this case were delayed treating and partial bark removal.   The bark was removed at numerous places on the bolts to simulate the result of rough handling; these barked areas were sprayed but were not treated with coating compound.   A single end-coating was used throughout; this was the asphalt cutback, which was among the most effective coatings tried on white birch bolts (table 3).

Except for a limited number of narrow streaks (penetrating as reported in table 4, under mean maximum depth), decay in the promptly treated bolts during 8 months of storage was slight.   This was true even with the bolts on which there were numerous barked areas.   By delaying the treating 2 months, appreciably more decay was permitted, but the average amount, nevertheless, was only about one-fourth that in the untreated bolts.

It should be noted that this comparatively successful outcome with delayed treatment is attributable to temperatures unfavorably low for fungus development during the 2 months (February and March) between cutting and treating. According to Weather Bureau records these temperatures averaged 18.6° F. in February and 27.5° in March, and on only

---

[5]See footnote 2.

2 days of this period were the average temperatures as high as 40°.

Suggestions For Treatments

Specific suggestions for chemical treatment of northern hardwood logs and bolts can be given most concisely in tabular form. The treating plan outlined in table 5 is based on test results, but necessarily includes many variables not actually covered by the tests. Because of this and the fact that there has been no significant industrial experience with such chemical treatment in northern hardwood areas, the plan is offered as a guide for commercial trials rather than as one to be adopted outright.

The question has been asked whether the fungicide might effectively be incorporated in the end-coating and thereby avoid the extra labor cost of applying it separately in a spray. Test evidence indicates that this is possible--provided the fungicide is one that is dispersible in the coating compound. However, more information about such a procedure is needed. For instance, in mild weather it probably will not permit as long a delay between cutting and treating as when the fungicide is applied in spray form.

Species Not Benefited

As pointed out, chemical treatment of the kind described is suitable for white and yellow birch and for hard and soft maples. Beech, however, is not appreciably benefited. Beech bark loosens rapidly and, in addition, seems to offer but small resistance to penetration by decay fungi.

In view of this over-all susceptibility to infection, unprotected beech logs and bolts might well be stored separately and utilized before the other species. They can be fully protected by water storage and, to a considerable extent, by storage under wet hay. Insofar as observed, storage under wet hay seems to bring beech through in about as good condition as other woods.[6]

Logs of black oak are likewise not much benefited by the described chemical treatment because of poor fungus resistance of the bark. It has not been ascertained whether this is true also for other northern oaks.

---

[6]Scheffer, Theodore C., and Zabel, Robert A. Storage of beech logs and bolts in the Northeast. Northeast. Forest Expt. Sta., Beech Util. Ser. 2. 13 pp., illus. 1951.

The cost of materials consumed in treating bolts can only be presented as approximations. For thorough spraying of ends, about 2 gallons of solution will be required per 100 square feet of surface. Using solution concentrations listed in table 1, this would entail a cost of approximately 10 cents. For the end-coating, about a gallon of material will cover 100 square feet. If asphaltic preparations of the kind tested are used, the cost for this would be about 50 cents. The total cost of chemical for both spraying and end-coating would be, then, approximately 60 cents per 100 square feet of treated surface.

Based on these estimates, chemicals used in spraying and end-coating a cord of turning bolts averaging 7 inches in diameter would cost about 25 cents. For similarly treating veneer logs averaging 14 inches in diameter and 16 feet long, the cost would be about 11 cents per 1,000 board feet log scale. To this, of course, labor costs would have to be added to determine total cost.

OTHER STORAGE DEFECTS

Wood-boring beetles seem to cause negligible damage to bolts and logs of northern hardwoods that are stored through no more than a single warm season. The decay infection introduced by these and bark beetles likewise appears to be of small importance.

At the end of a year's storage, a small amount of decay was found in wood just beneath the bark of the tested white birch bolts, and this plainly was introduced by bark beetles. It was very shallow in all cases. Similarly, in the yellow birch veneer bolts that were dry stored for 8 months, a small amount of decay occurred in the outermost sapwood as a result of infection introduced by roundheaded borers. Such decay was practically all eliminated in the process of rounding up the bolts on the lathe. Water storage for decay prevention is equally effective in preventing insect attack.

The dark-colored sap stain that is so common in logs and lumber of pine and of certain southern hardwoods is rarely seen in northern hardwoods. This type of stain is produced by fungi, and where it occurs it can be controlled by the same measures that are successful against decay.

Table 5.--Suggested treating plan for various months of cutting
and periods of storage

Month of cutting	Maximum delay in treating[1]	Last month of storage					
		May	June	July	August	September	October
November	4 weeks	C or D	C or D	C or D	C or D	C or D	C or D
December	6 weeks	C	C or D	C or D	C or D	C or D	C or D
January	6 weeks	C	C	C or D	C or D	C or D	C or D
February	6 weeks	C	C	C	C or D	C or D	C or D
March	4 weeks	C	C	C	C	C or D	C or D
April	1 week	B	B or C[3]	C	C	C	C or D
May	2 days	A	B	B or C[3]	C	C	C
June	2 days	--	A	B	B or C[3]	C	C
July[2]	2 days	--	--	A	B	B or C[3]	C

Key to treatments

A - No treatment needed.

B - Spray ends with a fungicide (table 1).

C - Spray ends with a fungicide and follow with a coat of end-sealer.

D - Same as C except apply 2 coats of end-sealer if maximum effectiveness is desired.

[1]For assurance of best results, treating at all times should be done shortly after cutting. The maximum delays of 1 week or longer are believed to be conservative, but nevertheless should be regarded only as a tentative guide.

[2]Material cut after July and used before the following spring probably would be adequately protected by treatment B.

[3]Treatment B for yellow birch and maples; treatment C for white birch. For material cut in June and to be held as long as 2½ to 3 months (e.g., early June to late August), it would be safest to use treatment C with all species.

Table 6.--Brown chemical stain in 15½-inch (mean diameter) yellow
birch veneer bolts during 8 months of water storage or
of dry storage with protection by spraying the ends
with 5-percent sodium pentachlorophenate and following with a coating of asphalt cutback[1]

Treatment	Bolts treated	Depth of stain from end surfaces at end of test	
		Mean maximum	Mean average
	Number	Inches	Inches
Stored in water	10	0.6	0.1
End-treated promptly after cutting	10	14.5	11.4
Same, except patches of bark were first removed to simulate effect of rough handling; barked areas sprayed but not coated	10	21.4	13.8
End-treated 2 months after cutting	10	20.4	14.1
None (controls)	10	20.2	18.4

[1]All bolts were cut in early February and the stain observations were made in early October. Decay in the same bolts is described in table 4.

An objectionable yellowish-brown to rust-colored
stain commonly occurs in birch logs during warm weather.
Like decay, it progresses inward from the ends and may pene-
trate rather deeply in a comparatively short time. It is
not of fungus origin, however, but is caused by oxidation of
certain of the wood constituents on exposure to the air. To
control it, air must be kept from entering the log ends.
This can be accomplished by storing in water. Test results
indicate that the asphaltic end-coatings may also be help-
ful, particularly in double thickness.

Quantitative appraisals of brown stain in the test on
yellow birch veneer bolts are shown in table 6. The single
coating reduced the average amounts of brown stain by approximately 30 percent. It is estimated that double coatings in the year-long test on white birch bolts (table 3)
were more than twice as effective as the single coatings
against brown stain.

SUMMARY

Logs and bolts of northern hardwoods can be stored
for long periods without significant decay if kept in fresh
water. Storage under a layer of wetted hay seems to give
moderate protection. Continuous wetting with a spray of
water has given variable results; it is sufficiently prom-
ising, however, to warrant further trials.

White and yellow birch and hard and soft maple can be
protected against decay during 2 or 3 months of summer
weather by suitably spraying the ends of the logs with a
fungus-inhibiting chemical solution. Protection over a much
longer period may be had by following the spray treatment
with a vapor-resistant end-coating that will prevent development of seasoning checks. Such a coating also shows
promise for reducing the yellowish-brown to rust-colored
chemical stain that occurs in stored birch logs during warm
weather.

A suggested plan for treating is given for different
months of cutting and for different periods of storage an-
ticipated. Logs of beech and black oak are not benefited by
the described chemical treatment because the bark of these
woods has little resistance to fungus.

THE MATERIALS USED

The chemicals and end-coatings mentioned in this report were obtained from the following sources:

Dowicide G: Chapman Chemical Co., Memphis, Tenn.

Permatox 10-S: Chapman Chemical Co.

Melsan: E. I. du Pont de Nemours and Co., Wilmington, Del.

Noxtane: Wood Treating Chemicals Co., St. Louis, Mo.

Santobrite: Monsanto Chemical Co., St. Louis, Mo.

Asphaltic end-coatings (C-13-HPC, C-13-E, Semi-Mastic 214): Flintkote Co., Chicago Heights, Ill.

No attempt has been made to draw up a complete list of manufacturers or distributors of such materials. This list is not an endorsement of the quality or price of the products mentioned; and no discrimination is implied against firms whose products have not been mentioned. Products of other manufacturers are obtainable that will undoubtedly serve the purpose as well.

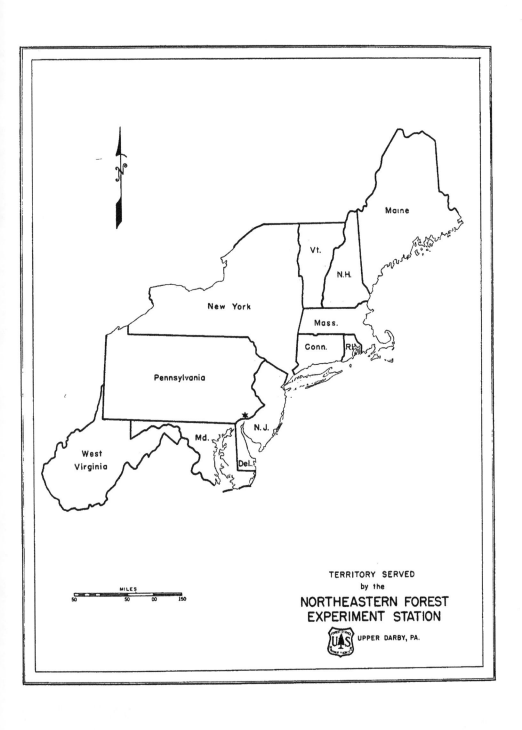

TERRITORY SERVED
by the
**NORTHEASTERN FOREST
EXPERIMENT STATION**
UPPER DARBY, PA.

Lightning Source UK Ltd.
Milton Keynes UK
UKHW012004021218
333216UK00014B/2701/P